A Field Guide to
BLUEBERRIES

A FIELD GUIDE TO
Blueberries

Jim Johnson

North Star Press of St. Cloud, Inc.

Text illustrations: Dale Hagen
Author photo: Marlene Wisuri
Cover art: Jessica Allen Johnson

Library of Congress Cataloging-in-Publication Data
Johnson, Jim, 1947-
   A field guide to blueberries / Jim Johnson.
   64 p. 21½ cm.
   ISBN: 0-87839-072-3 (pbk.) : $9.95
   1. Blueberries—Poetry. 2. Minnesota—Poetry.
I. Title.
PS3560.037917F5   1992   92-18805
811'.54—dc20   CIP

Printed in the United States of America by Versa Press, Inc., East Peoria, Illinois.

Published by North Star Press of St. Cloud, Inc., P.O. Box 451, St. Cloud, Minnesota 56302.

ISBN: 0-87839-072-3

Table of Contents

A Field Guide to
BLUEBERRIES

Ask for Buckets, You Get Cupfuls

I give to you:
      small trout from unnamed brooks
      mushrooms
      full moons
      the contents of a grouse's craw:
            51 clover leaves
            34 alder cones
             9 twigs with deerfoot buds
             1 piece gravel quartzite
and blueberries.

When Picking Blueberries
An Early Start Is of Utmost Importance

In Lake County along the road
in a slough
in the early morning still silted with the
still of night, a mist coming off
when and a moose—
humped darkness that arced into dark water
(as I stopped the pick-up truck) suddenly
rose
out of the slough
upslurping all silence—an antlered head huge,
great clumps of distant timber remembered,
earth clinging to the roots, its eyes
large
bullheadedly round, not yet of daylight,
and
its mouth nonchalantly moving stems to
dangle,
lilypads already turned lavender.

In Blueberry Country

Sign by the road: *Minnows Crawlers Leeches.*
I remember: fish bite best when the mermaid's tail
turns blue.

The Names Loggers Give to the Trees
Often Differ

Because loggers built their own roads
the law didn't apply to them.
They brought their children up
believing. I remember when
I drove a narrow
gravel road and met
a logging truck
doing 50
no room to get
over. Somehow
I got by.
The road rose.
The sunlight slanted
between the pines.
Dust filled everything.
I thought of having come
through one life
and into another.
So many questions. Now
the stones
no longer listen.

Heading North

Through the pine shadow
a crow flies
parallel the dusty road.

Where They Grow, How Seriously Taken

In the clearing cleared by loggers, blueberries
acres and acres of
blueberries, each one a different color.

When to Pick

It's best to pick on a clear day with some wind,
    a high blue sky.
Still humid days are the worst.
Even on a dry day black flies buzz your hands,
              buzz your arms,
              buzz your face, any exposed skin.
As you pick you brush your face,
          brush your arms,
          brush your hands between every bunch you
pick, picking up a life, if you must know, is not your life. Then
and only then, if you must not, it is.

There Are Several Methods

Some pick clean
one at a time, others fast,
bunches at a time, which is best in open areas.
Usually the low bush
ripens first. Near the lob pines
where the berries are bigger, but only
one berry
is ripe and three or four white per bunch,
one at a
time is best, as in the lowlands among the moss
where I have seen berries even into the end of
September
with a morning cap of snow.

*Blackflies—Small Humpback Gnats Not Confined to
the North Country, Breed in Flowing Water, Lay Eggs
on Sticks, Rocks Projecting into Water, Larvae
Clinging to Objects (Old Shoes, Beer Cans)
in Water, from Which They Gather Food with Set of
Motile Brushes around Mouth, Spin Web-like Pockets
Underwater, and Pupate—Now*

About your neck
      your ears
like an old woman with knitting and questions,
occasionally   one gets in.

In Church I Counted the Light Fixtures
So I Couldn't Tell You Where Is the Hell
Preachers Tell, Only That

Hot and humid. Blackflies bad.
Headnet on,
the world drab green
no holes in the ozone
yet
somehow they get in
inside the headnet
inside the head. On the outside you can't tell
which are in and which are
out, only that there are millions
millions swarming,
and as you smash one and another and another
there are no less.

What Can Only Be Reached through Much Suffering

Put your hand behind your ear, feel
small and squished, what you know is
black,
look how the blood divines your finger.

How Blueberries Are Measured

At home
the small red dots along your arms, behind your ears,
above your sockline, your waistline, in the corner of
your eye—small red dots where blackflies got in,
        small red dots about the same size

                      as blueberries
not quite ripe.

The Leaf Blade Is Dual-Shaped and Entire

A bucketful of leaves, spruce needles, twigs, reindeer
    moss, worms the size and shape of blueberry stems,
    and blueberries,
each one a different color of dark,
each one a different phase of the moon,
each one so blue, blue beyond the edge of believing
(even Columbus knew north varied four degrees)
plunked into this galaxy of stainless steel.

When the Wood Cells Die, They Leave
a Network of Vertical Bundles

On a stump cut years ago and dark stained
with a pattern of growth rings, a green leaf
with dried edges   like a grasshopper
lives again.

While Blueberries Are Blueblack and
Often Covered with a Light Blue Bloom

Rosehips are another red like brook trout eggs
in August when neither are ripe.

Bluer Than a Bottle of Great Doctor Kilmer's
Swamp Root, Blacker Than the Particles of
Night, Brighter Than Bear's Hair, Cubs Hung
from Its Summer

The true blueberry is
heaven and earth,
           the four winds and dark water,
yet it tastes
of nothing.

Fruits Developed from the Ripened Ovary
Bear Seeds by Which Plants Reproduce

In the open clearing logged over years ago
blueberries
overripe, dried, and shriveled up.
Even as I pick many fall to the ground
or go unpicked. In each
hundreds of seeds. At the Co-op in Finland Minnesota
last March
I thought I saw old Väinämöinen
in a black snowmobile suit, an Arctic Cat hat.
He was feeling up the tomatoes.

Clintonia

False blueberries I call them,
a single berry
like a blue lightbulb on a stem
growing out of three lily-like leaves,
growing so near
blueberries and looking so like
blueberries I always find a few
in my bucket after picking.
They taste bitter.   Maybe poisonous,
maybe not.   The true with the false, the false with the
true.   Without order and in reverse
order.   Hidden as much as possible.   A separateness
beginning with a name.

On the Naming of Names

Far to the north
summer is an illusion
we live on. Once at the end of a long winter
a day so warm
water flowed back into the hole
cut in the ice. It was a lake
with no name. On the map
another lake with a stream flowing in
and a name. It looked like a trap and chain.

The Taste Is Everything

How would it be to have a blueberry contest
to see who could bring in the biggest blueberry?
You couldn't.
Someone would bring in a domestic blueberry,
say it was wild,
and you couldn't tell it wasn't,
unless,
of course, you tasted it.

At One Time the Intent of Poetry Was
to Instruct

A Co-op coffee can makes a good bucket
if you drill two holes near the top
and fit them with any thick wire for a bail.
Be careful not to get the bucket wet or
the bottom will rust. If it does, it is
not much good. Use it, then, for picking rocks.

The Logic of Flowers

This is the forest that loves its purple
flowers and neglects
its yellows. No matter how prodigious
they will always be common. In areas
struck by loggers, lightning, and
left out in the open
they plant their flags of conquest
in the uncharted
soil. They spread the word.
It is not enough.
This is the forest that knows only
the logic of flowers.

*The Trouble with Great Wisdom and Local
Poets*

Morning hot and humid,
right off. Birds talked,
said it would rain. Before long
clouds blotted the sky, the sky solidified
to gray,
and we felt raindrops, ran
to the truck, and still doubted.

Looking for an Edge Where There Are Many
Edges, a Tree Where There Are Many Trees,
a Road Where There Are Many Roads, a Word
Where There Are Many Words

Sometimes all that is found is an empty
bucket or hat.
People do get lost. I call or wave.
Just a rock.
Just a log, an overturned log
roots in the air
like antlers.

And the White Pines Grow Old as the Sky

You sought out the lonely places. The road in
was very narrow,
many rocks, ruts, chuckholes,
the road so bad you wanted to stay in
second gear
and close your eyes. Instead
you got out and walked. The roads gave way
to trails.
Worms bored inside the bark of trees.
When he worked with wood
      he looked to the wind
            the waves
            the snow drifting. But
you did not want to be the wife of an old man.
You wept. All around you was a swamp.

Arguments against a Blueberry God

I once brought berries and small trout to an old
woman. She made me coffee, she herself was full of
grounds, and taught me the words to a language I
had long forgotten. Then one day she paused, and
her face fell. First it shifted. The ear, cheekbones,
mouth on one side faulted as if in half a frown. Then
it all fell, fell to the floor. It had such suction that,
try as she did, she could not pick it up. Her hands
were wet with blood. Her fingernails too short. She
could not pick a corner up. To make matters worse,
the face began to cry. The river I fish in is now a
broken mirror.

Because as Kids We Picked with Snoose
In Our Mouths and Brought Them All Home

Some people stopped to tell me not to pick,
they saw
a bear nearby. Illinois license plates.
I went on picking. Then
it was overcast. It was winter. A popple
felled against a mound of snow. The mound shook.
Leaves, roots frozen to my black fur,
I emerged
a bear! into the daybright, what was before me,
two-legged and with an ax—I turned and ran,
ran until I realized
my shadow did not breathe down on me
the hot breath of death, but instead
was unhinged,
running across the opposite snow. Perhaps
it too would turn
realizing how much alike we really were.

In Isabella

They cut timber in Isabella, Minnesota.
Sell gas, bait. Minnows only.
No leeches or ice. Work on
skidders. Saturdays off.
Come Monday it takes a long
time to spit snoose, rub it into
the ground, look at a
chainsaw. How the blade
is shaped like a beaver's tail.
Around the blade are teeth.
The body usually a color like
yellow, but about the same
size. To skin it out
simply unscrew a nut behind
the handle. Around here
some people think they are a nuisance.
Cutting down the trees, damming
the streams. Where the meadow grass
narrows, a point of firs.
The waters backed up like twilight
are the stained glass windows
of the log cathedral. Artisans
worked for years piecing together
pieces of Jesus feeding the multitudes
loaves and fishes. Men will lay down
a foot or leg before a chainsaw.
Beaver men lived their lives for
someone else's hat.

The Bark: Dark, Thick, Deeply-Furrowed
with Broad Ridges Composed of Closely
Pressed Purple-Tinged Scales

Like the winter-hardened faces
that go out onto the ice in early November
when the ice is thin
and clear, no snow, so clear
maybe they see the faces of the children,
the ones who died, the ones who were too bright.
Now they are the ones who cut down these pines.
You never see them weep
or look up too long
before they get out of the way.   The Indians used
the inner bark for food, an ingredient in cough syrup.

Historical Note: They Came Because the
Winters Were Long, the Summers Short

Along a road where many cars trailering boats go by
        bunches of large ripe dusty black even blacker
        than blue blueberries
no one stops to pick. Now you know
what the old country people know. An abandoned house
in the middle of a clearing
miles from any road. A place so quiet
you can leave and watch yourself
go on picking.
                What do you believe?
                I believe in an old man.
                An old man who is god?
                I believe in an old man who is god.
                He came from the old country
                a long time ago.
                Now the new world is no longer new.
                So what are you going to do?
                I am going to go on.

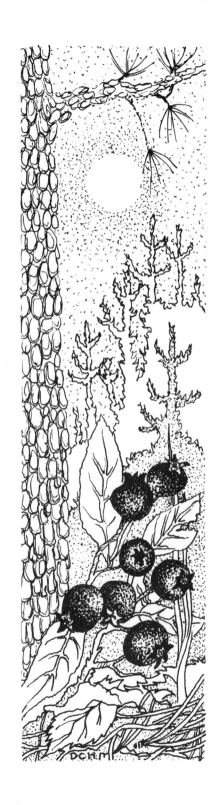

Another Step Forward

Walking through the brush once logged over,
now grown up with waist-high cover
I hear something ahead
—I stop. Then silence.
I go on,
hear it again. Again stop. And again
silence. I go on
as the other end of the fir log
I am stepping on
takes another step forward.

Sometimes the True Self Is Revealed

Difficult as it is to accept after walking so far
in what you thought was one
direction
except for stepping over windfalls,
        raspberry thickets, or low spots, but each
time correcting the approximate course only
            to find
after walking miles of not finding
the compass needle
       blue and wavering,
wavering as if it were your own soul, spinning
around and around
until it settles on a course—
the wrong course!   You insist upon finding out
it does not point in the direction you have been
going.   Of course
the compass needle must be wrong,
the curve of its own
       variation,
           narrower and narrower,
now searching for, if level and no iron
     in the ground, what is always
        true inside.

The Way with Everything

There is a way with everything.   Like the course
a stream follows.   Once a moose skull lay
beside a stump
along a section line.   The way I chose was
not the section line but through the spruces
                        around windfallen logs thickets
picking a way. Drawn to that skull
was another mind like mine.

Wear Sturdy Shoes, Long-Sleeved Shirt
Buttoned to the Top, Long Pants, and a
Hat, and Do Not Oil Boots with the Oil
of Prey

Close to the road, dusty and low to the ground
              and hard to find, the stems tangled with
              sticks, dried grass, and
for a long time see nothing, nothing to pick, and
now
now having the time to think
now having the time to think
turn over and over
                        another bush.

The Conifers Are Identified by Needle-like Leaves

In the slanting light between the trees
a fir branch moving, moving like a bird's wing.
It is a bird's wing.

Sometimes
the Wind Picks Up on a Minor Detail

Something moving:
two leaves, two yellow popple leaves
on a hinge: a butterfly.

Seeing

No wind.
In the clearing
cleared by loggers
a hawk perched
on a dead fir
the top broken off
seeing only movement,
seeing being
the only movement.

Inside Each Berry a Thousand Seeds
Like the Ends of Popple Logs Cut and
Stacked, the Pith Developing Like a
Fetus, in Each Log
                    A Different Shape

The wolf scat that I found in March
on the trail of crusted snow
was nothing more than
excreted brown fur.   A few days
a south wind
and the snow hardened by premature melt,
the scat would dry out
whiten
and come apart,
each individual hair
seeping into the ground
                    like roots.

All Part of the Community Where
Plants and Animals Live

Pull it through your fingers.
It is long and thin,
a red fox hair,
with the slightest bit of
fox skin at one end
and tapering to nothing
at the other.
Pull it across your nostrils.
Smell the fox within.
Feel your back begin to hackle.
Paws begin to twitch,
something wild,
something sniffing sniffing
sniffing for more. . . .

The Way a Tree Is Hinged, It Creaks

By the blacktop, in the box of a pickup truck
a Dodge Powerwagon
another open box made of plywood
blackened with grease (blacker than the stories
Lutherans told of Catholics).   In it
a chainsaw
a tank of gas
a couple of cans of chain oil
a couple of wrenches
and no one in sight. Just gravel dried leaves
and pollen stuck
to the blackened box of plywood
made by a logger who thinks lord knows what
of man and woman and a meadow
          heaven and hell
                    being a good neighbor.

At the Knotted Pine

A battered Chevy pickup truck
parked outside.
Inside
at the bar Ollie Thoms
aged 65,
10 kids and 10 fingers,
drinks down a Pabst. *Nowadays,*
he says to me after I buy him another,
If you want to go a-logging
you need your own machinery.
You need your own machinery.
First used chainsaws up here
around '58. Now
you wouldn't want to do
what we used to do
without one. You wouldn't
want to fall on one
either.

Myriads of Root Hairs Extend into the
Soil Increasing the Absorption Surface
Tremendously

As I look out at the lake I remember you and me
swimming
a loon
black and white arcing wingspan
suddenly in a splitting of a beak
instant dripping
surface to surface
this moment so near to us
that
the pump out back, the water cool and clean
from forty feet deep, the blueberries in season,
small trout, cattails, tall white pines
that all these years unlogged kept the cabin
cool in summer
all made perfect sense.

As Identified by Soft Blue Needles Five
to a Bunch

When I touch the ends of your hairs with my tongue
it might be the wind licking the treetops.

Priming the Pump

I pour water into the head of the one-armed fugitive
       from the assembly lines of Butler, Indiana;
I'm glad to have him here after all these years
       hauling water. Now when
I pour more water and stroke the only arm,
the pump coughs.
I pour more water,
       pumping the water through, pumping, pumping
until the water flows from the mouth—
water that has never seen the light of day
water coming from the throat, first brown water,
water pouring out, so much
water pouring out,
water running over my bucket, the bucket of one
       who has suddenly found the words to say
what is now pure and clean, from deep beneath
what is and will ever be
       from the heart.

In the Blueberry Bucket

A man picks blueberries all day. In the bucket
small green and withered sifting down,
too old and squashed against the sides,
twigs leaves spruce needles settled to the bottom
while the large ripe rise up to the top.
So full of flesh, so ordinary—
a woman, at sunset
a blond woman leading a reindeer
across the horizon. You are that woman. I am that man.
And this I give to you
is, if not the history of the world
at the end of the day, but
a bucketful of blueberries.

In Isabella, Minnesota, I Have Found Crystals
of Dirty Snow under a Log, Even in July

And blueberries
        green blueberries, red blueberries, ripe blueberries
        blue blueberries, blueberries blue as Finland
        blue-black blueberries, black-blue blueberries
        white-spotted blueberries, withered white
        blueberries, thousands and thousands of
        blueberries—each one the end of summer.

Driving Home while the Moon Repeats Itself
Not Far from Isabella, Minnesota

The August moon above the trees
not quite full
like a Canadian dollar, in and of itself,
shining down on a last tangent of lilypads.

Why There Are No More Songs About the Moon

Once the blueberries were few.
Each year they were fewer and fewer.
You said we wouldn't find any and
I said
      I knew a place.
At first we didn't. Then
near an edge
a few. Then a few more. Then a bush
and several more each the size of a TOMATO.
From that bush I filled my bucket. While
the berries were large
they were few. Each year they were
fewer and fewer
and didn't
      taste as sweet. The moon too
is a solitary blueberry
not quite ripe. The astronauts that walked on it
not even the size of ants. No rain in the forecast.

Afterwards

Afterwards in the kitchen
with two buckets of blueberries,
I drink wine from a water glass
and feel
behind my ears, along my hairline.
I close my eyes
and see,
see only blueberries,
and know
there is a wisdom in all things,
                    except mosquitoes.

Recipe for a Blueberry Pie

No one else can pick them for you. So you must put all—homegrown tomatoes, the chance of frost, more stovewood—aside and go. Suffer the blackflies and the mosquitoes. Go out into the woods. Once near Bear River I actually saw a bear. It ran across the highway. Only later, much later, did I come to believe it was, it ever will be, and that that is why we make pies.

Jim Johnson is a Cloquet, Minnesota, native who teaches in Duluth, Minnesota, and writes poetry that reflects his northern Minnesota and Finnish heritage.